U.S. Department
of Transportation
**Federal Transit
Administration**

ELECTRONIC FARE COLLECTION OPTIONS FOR COMMUTER RAILROADS

FINAL REPORT

September 2009

Report No. FTA-MA-26-7109.2009.01

U.S. Department of Transportation
Federal Transit Administration

ELECTRONIC FARE COLLECTION OPTIONS FOR COMMUTER RAILROADS
September 2009

Prepared by:
Lydia Rainville, Victoria Hsu, Sean Peirce
U.S. Department of Transportation
Research and Innovative Technology Administration
John A. Volpe National Transportation Systems Center
Economic and Industry Analysis Division
Cambridge, MA 02142

FINAL REPORT
Report Number: FTA-MA-26-7109-2009.01

Sponsored by:
Federal Transit Administration
Office of Research, Demonstration and Innovation
U.S. Department of Transportation
1200 New Jersey Avenue, SE
Washington, D.C. 20590

Available Online
[http://www.fta.dot.gov/research]

Report Documentation Page			*Form Approved* *OMB No. 0704-0188*
Public reporting burden for his collection of information is estimated to average 1 hour per response, including the time for reviewing instructions, searching existing data sources, gathering and maintaining the data needed, and completing and reviewing the collection of informa ion. Send comments regarding this burden estimate or any other aspect of his collection of information, including suggestions for reducing this burden, to Washington Headquarters Services, Directorate for Information Operations and Reports, 1215 Jefferson Davis Highway, Suite 1204, Arlington, VA 22202-4302, and to the Office of Management and Budget, Paperwork Reduc ion Project (0704-0188), Washington, DC 20503.			
1. AGENCY USE ONLY (Leave blank)	2. REPORT DATE September 2009	3. REPORT TYPE AND DATES COVERED Final Report September 2008-September 2009	
4. TITLE AND SUBTITLE Electronic Fare Collection Options for Commuter Railroads		5. FUNDING NUMBERS MA-26-7109	
6. AUTHOR(S) Rainville, Lydia; Hsu, Victoria; Peirce, Sean			
7. PERFORMING ORGANIZATION NAME(S) AND ADDRESS(ES) U.S. Department of Transportation Research and Innovative Technology Administration John A. Volpe National Transportation Systems Center Economic and Industry Analysis Division Cambridge, MA 02142		8. PERFORMING ORGANIZATION REPORT NUMBER	
9. SPONSORING/MONITORING AGENCY NAME(S) AND ADDRESS(ES) U.S. Department of Transportation Federal Transit Administration 1200 New Jersey Avenue, SE Washington, DC 20590		10. SPONSORING/MONITORING AGENCY REPORT NUMBER FTA-MA-26-7109-2009.01	
11. SUPPLEMENTARY NOTES Available Online [http://www.fta.dot.gov/research]			
12a. DISTRIBUTION/AVAILABILITY STATEMENT This document is available to the public from: The National Technical Information Service, Springfield, VA 22161. [http://www.ntis.gov]		12b. DISTRIBUTION CODE TRI	
13. ABSTRACT (Maximum 200 words) This research is designed to support FTA in its efforts to disseminate knowledge of new technologies within the transit community, in this case focusing on issues associated with automated fare collection (AFC) for commuter rail. By identifying "lessons learned" with AFC deployments, the report is also intended as a resource for commuter railroads considering adoption of AFC and/or joining multi-modal electronic payment systems. The findings may be of particular use for commuter rail systems that are still in the planning stages and have the opportunity to leapfrog older technologies.			
14. SUBJECT TERMS Transit fare payment, commuter rail, electronic fare collection, electronic payment system, AFC		15. NUMBER OF PAGES 44	
		16. PRICE CODE	
17. SECURITY CLASSIFICATION OF REPORT Unclassified	18. SECURITY CLASSIFICATION OF THIS PAGE Unclassified	19. SECURITY CLASSIFICATION OF ABSTRACT Unclassified	20. LIMITATION OF ABSTRACT

NSN 7540-01-280-5500

Standard Form 298 (Rev. 2-89)
Prescribed by ANSI Std. 239-18
298-102

FOREWORD

This report discusses technical and institutional issues and "lessons learned" related to the implementation of automated fare collection systems on commuter railroads. It was sponsored by the Federal Transit Administration as part of its efforts to disseminate knowledge related to new technologies and emerging trends within the transit community.

DISCLAIMER NOTICE

This document is disseminated under the sponsorship of the Department of Transportation in the interest of information exchange. The United States Government assumes no liability for its contents or use thereof.

The U.S. Government does not endorse products of manufacturers. Trademarks or manufacturers' names appear in the document only because they are essential to the objective of this document.

ACKNOWLEDGMENT

We would like to thank the commuter rail agencies for their assistance and cooperation in helping us prepare the case studies. All photographs are reproduced with the permission of the railroads.

Table of Contents

Table of Contents	ix
List of Tables	ix
Executive Summary	xi
1. Introduction	1
1.1 Objective	1
1.2 Methodology	2
2. Scan of Current Practice	3
2.1 Background on Fare Collection	3
3. Case Studies	10
3.1 Case Study Summaries	10
3.1.1 New Mexico Rail Runner Express	10
3.1.2 Virginia Railway Express	12
3.1.3 Sounder Commuter Rail	14
3.1.4 San Diego Coaster	17
3.1.5 Metropolitan Transit Authority New York - Metro-North	19
3.1.6 Shore Line East	21
3.2 Lessons Learned	22
4. Domestic and International Experience with Automated Fare Collection	24
4.1 Closed-Loop (Proprietary) Systems	24
4.2 Open-Loop Payment Systems	25
4.3 Future Outlook	26
5. Conclusion	29
Bibliography	33

List of Tables

Table 1: Commuter Rail Fare Collection—Current U.S. Practice	5

Executive Summary

Over the past two decades, urban transit agencies in the United States and abroad have moved toward various forms of electronic payment systems (EPS) and automated fare collection (AFC), often magnetic stripe cards or smart cards. These developments have been driven by benefits ranging from reductions in the accounting and cash management costs to opportunities for enhanced data collection to support planning and operations. However, applying these technologies to the open, barrier-free layouts of commuter rail stations has proven challenging, and adoption has lagged behind that of other transit modes.

The six case studies in this report provide some insight into commuter railroads' experiences in adoption of AFC. Currently, only Central Puget Sound Regional Transit Authority and North County Transit District have implemented a smart card that will include their commuter rail services. Two other agencies, New Mexico Rail Runner and Metro-North, are using handheld devices to facilitate onboard ticket sales.

The lessons learned from these case studies are intended to be of use to agencies considering a move to AFC. For proposed commuter rail systems that are still in the planning stages and can thus implement a new fare-collection system from scratch, this information can also help to identify the pros and cons of different approaches

Additional insight on electronic fare collection for transit and some indications of future directions can be found in the experiences of other agencies around the world, though most innovations have been in urban transit rather than in commuter rail. London and Hong Kong both have years of experience with successful, proprietary closed-loop systems and co-branded cards. London is now looking to expand to an open-loop system. New York, Paris, Malaysia, India, and Turkey have also conducted pilot tests. As new technologies emerge, whether in commuter rail or other types of transportation, ongoing study will help provide a sense of their performance and suitability for the commuter rail environment.

1. Introduction

1.1 Objective

Over the past two decades, transit agencies in the United States and abroad have moved toward various forms of electronic payment systems (EPS) and automated fare collection (AFC), often magnetic stripe cards or smart cards. These developments have been driven by the wide range of advantages that AFC enjoys over traditional paper tickets and manual fare collection, including:

- Improved throughput
- Reductions in the costs of cash management, record-keeping, and currency counting
- Reduction in fare evasion
- Lower potential for "revenue shrinkage"
- Interoperability between regional transit fare and transit parking payment
- Stored-value systems, which provide additional convenience for customers
- Opportunities to introduce loyalty programs/flexible discount structures
- Improved customer satisfaction and reduced costs, introducing flexibility to both operators and customers
- Improved labor relations with conductors if burdensome procedures or equipment can be eliminated
- Safety and security benefits, such as more accurate passenger counts in the event of an emergency, and reductions in the amount of cash that conductors need to carry
- Opportunities for enhanced data collection capability to support transit operations and planning – notably origin-destination passenger counts based on AFC use at boarding and alighting
- Opportunities for the AFC technology to support other uses, such as two-way communication and transit provider credentials
- Increased collection of revenues prior to service, leading to realization of revenue from "unused" transit value

Despite this range of benefits, adoption of AFC on commuter rail systems has proven challenging and has lagged behind that of other transit modes. The open, barrier-free layout of many commuter rail stations makes it difficult to implement automated payment approaches. Given the long distances involved, most commuter rail systems also use zone- or distance-based fare structures rather than a flat fare, which introduces complexities in charging each rider the appropriate fare. Commuter railroads also tend to have a large proportion of riders who are regular commuters and use monthly or multi-ride passes. For occasional riders, conductors sell single-ride and round-trip tickets on board.

Just in the past year, a few commuter railroads have begun adopting EPS, for example by introducing handheld devices that can be used by conductors, or by participating in the development of a regional transit smart card. In addition, new partnerships are emerging between transit agencies, financial institutions, mobile communications companies, and transaction processors. These developments present many opportunities to improve EPS on commuter rail systems, yet most agencies have only limited information on the latest innovations.

This research is designed to support FTA in its efforts to disseminate knowledge of new technologies within the transit community, in this case focusing on issues associated with AFC for commuter rail. By identifying "lessons learned" with EPS deployments, the report is also intended as a resource for commuter railroads considering adoption of AFC and/or joining multi-modal electronic payment systems. The findings may be of particular use for commuter rail systems that are still in the information-gathering or planning stages and have the opportunity to align with other transit modes.

1.2 Methodology

This report covers Phase I of an envisioned two-phase effort. As noted above, it is intended to provide a general overview of commuter railroad experience with EPS as well as key lessons learned. (Phase II, if pursued, will build on these results with more in-depth analysis of the financial, management, and technology issues associated with specific EPS deployments, as well as optional approaches and utilization of the most cutting-edge applications of technologies.) Findings in this report are based on a literature review, technology scan, and case-study telephone interviews with six commuter railroads. The case studies were selected to cover a variety of experiences with AFC – ranging from those who have deployed full-fledged systems to those who have not pursued AFC at all, and some in between – and to reflect a balance between large and small agencies, different regions of the country, and different underlying approaches to fare collection.

As a starting point for the analysis, Section 2 of this report provides a high-level summary of fare collection practices at all of the country's commuter railroads. This overview provides a snapshot of the current state of the practice, and allows the case-study material to be viewed in context. Section 3 of the report summarizes the findings from the six case studies, with an emphasis on experiences in the financial, technological, institutional, operational, and customer service areas that may be of relevance to other commuter rail agencies. Section 4 presents a broader discussion of other AFC approaches that have been developed in the U.S., plus some international experiences and examples of other emerging payment technologies. Section 5 summarizes these findings and identifies areas for further research.

2. Scan of Current Practice

This section presents a high-level overview of fare collection practices at each of the country's commuter railroads. This information is intended to inform the reader's understanding of the current state of practice in this area, and thus enable review of the case study findings in that context.

2.1 Background on Fare Collection

A more complete discussion of fare payment and collection options and technologies is beyond the scope of this work, but can be found in related literature, including several Transit Cooperative Research Program reports.[1] However, several dimensions of fare collection are worth mentioning in brief, as is the historical progression in the approaches that have been used in U.S. transit systems.

Fare structure can take several forms: a flat fare for travel between all points, distance-based fares that reflect the distance between origin and destination, or zone-based fares that are calculated based on the number of defined geographic zones through which the passenger travels. Fare structure also includes the various discounts that may be offered for seniors, students, or other groups. A second dimension is the **fare type** purchased. These typically include individual one-way and round-trip fares, multi-rides, weekly or monthly passes good for unlimited travel during a particular time period, and stored value passes that debit the customer's account with each ride.

Fare media also vary widely. Many of the earliest American transit systems simply used *coins* (often nickels and dimes) to collect fares, but over time moved toward selling *tokens* with a specific fare value. These systems are relatively simple and do not require much advanced technology, but they involve high costs for transporting, counting, and auditing the coins, currency, and tokens. They are rarely used for commuter rail because they are not suited to ungated environments, distance-based fare structures, or proof-of-payment approaches. For systems employing onboard inspectors or conductors, *paper tickets* and *passes* represent a low-tech approach with low upfront costs which allows quick collection and verification. In some cases they can be made compatible for use on other connecting transit systems, for example with special stickers or endorsements. While efficient, these

---

[1] Relevant TCRP reports include: *TCRP Report 10: Fare Policies, Structures, and Technologies* (1996); *TCRP Report 32: Multipurpose Transit Payment Media* (1998); *TCRP Report 80: A Toolkit for Self-Service, Barrier-Free Fare Collection* (2002); and *TCRP Report 94: Fare Policies, Structures, and Technologies: Update* (2003).

approaches require transit providers to establish an agreed-upon formula to divide revenues. Paper tickets still involve significant cash management and auditing costs since they are essentially bearer instruments with a cash value. They also provide only limited data on origins and destinations and on ridership as distinct from revenue.

Although paper tickets and passes remain common in commuter rail, many urban transit systems have long since moved to *magnetic stripe cards*, which are relatively inexpensive technologies that provide some additional security and data collection options, and are compatible with distance-based and variable fares. These cards do, however, wear out after repeated contact with electronic readers, and some types can also become demagnetized in wallets. More recently (in the past five to ten years), many agencies have adopted *contactless smart cards*, sometimes as part of a regional consortium. While more expensive to produce, these cards are longer-lived, and offer faster transaction times, greater customer convenience, and more opportunities for fare integration across agencies. Generally these smart cards are part of a proprietary closed-loop system, where the transit agency performs card issuance and account maintenance functions. Acceptance is limited to transit services and any transit-oriented retail sales, such as newsstands located on the transit providers' property, which are members of the closed system. By allowing customers to put large sums on the cards (some with "balance protection" in case of loss), they can also help to reduce the amount of cash in the system that needs to be tracked and audited by reducing the small bills collected. Finally, some agencies have considered accepting *bankcards* at the turnstile (an open-loop system). For customers without bankcards, prepaid cards issued by a third party, and linked to an external account, are also being targeted. This bankcard approach introduces new challenges about equipment needs, transaction fees, data security, and branding. The approach also has a variant: "co-branded" bankcards, on which a transit application and a bank/debit application are co-resident on a financial-institution-issued card.

As described in greater detail in Section 4, electronic fare collection approaches also create opportunities to partner with the private sector on system approaches, such as "co-branded" and "dual-interface" cards. Examples would include a debit/ATM card with a separate stored-value account for transit fares, or an employee ID card, which doubles as a transit pass. In some cases, these cards incorporate more than one interface technology so that they will work with legacy systems or in different environments.

Electronic Fare Collection Options for Commuter Railroads – Final Report

Table 1: Commuter Rail Fare Collection—Current U.S. Practice[2]

Commuter Rail / Transit Agency	Weekday Ridership (000s) (2)	Fare Media	Fare Options	Fare Structure	Peak Fares	Onboard with Surcharge	Onboard No Surcharge	Onboard Credit Card	No Onboard Sales	TVM	Ticket Window	Web/Mail	3rd Party Vendor
MTA Long Island Rail Road	350.0	P	S, M, T	Z									
Metra - Chicago, IL	335.9	P	S, M, T	Z	x	x				x	x	x	
Metro-North - New York, NY	292.8	P	S, M, T	Z		x				x	x	x	
New Jersey Transit	276.0	P	S, M, T	D	x	x				x	x		
Mass. Bay Transp. Authority (MBTA)	144.1	P	S, M, T	Z		x				x	x	x	
Southeastern PA Transportation Authority (SEPTA)	127.3	P	S, M, T	Z	x	x				x	x	x	x
Metrolink - Los Angeles, CA	47.6	P	S, M, T	D	x				x	x	x	x	
Caltrain - San Francisco, CA	44.9	P	S, M, T	Z					x	x	x	x	
MARC Train - Baltimore, MD	33.1	P	S, M, T	D		x				x		x	x
Virginia Railway Express (VRE)	15.6	P	S, M, T	Z					x	x		x	x
Tri-Rail - Miami, FL	15.3	P	S, M, T	Z	x				x	x	x		

[2] Fare Media: P=Paper Ticket, SC=Smart Card; Fare Options: S= Single-Trip, M-Multi-Trip, T=Time-Based; Fare Structure: D=Distance-Based, Z=Zone-Based

September 2009

Table 1: Commuter Rail Fare Collection—Current U.S. Practice[2]

Commuter Rail / Transit Agency	Weekday Ridership (000s) (2)	Fare Media	Fare Options	Fare Structure	Peak Fares	Onboard with Surcharge	Onboard No Surcharge	Onboard Credit Card	No Onboard Sales	TVM	Ticket Window	Web/Mail	3rd Party Vendor
Northern Indiana Commuter Transp. District	15.0	P	S, M, T	Z		x					x	x	
Trinity Railway Express - Dallas/Ft Worth, TX	10.9	P	S, T						x	x		x	x
Sounder - Seattle, WA	10.4	S, P	S, T	D					x	x	x	x	
UT Transit Authority - FrontRunner	7.9	P	S, T	D					x	x	x	x	x
Coaster - San Diego, CA	7.1	S, P	S, T	Z					x	x	x	x	
Capitol Corridor - Sacramento Amtrak Service	6.0	P	S, M, T	D		x				x	x	x	x
Altamont Commuter Express - Stockton, CA	3.7	P	S, M, T	D			mid-day train only		x	only to validate	x	x	
Alaska Railroad	3.5	P	S, M	D	x				x			x	
NM Rail Runner Express	2.7	P	S, M, T	Z			x	x				x	
Shore Line East - New Haven, CT	2.3	P	S, M, T	D			x				x	x	
PA Dept. of Transp. Amtrak - Harrisburg, PA	1.7	P	S, M, T	D					x	x	x	x	

September 2009

Table 1: Commuter Rail Fare Collection—Current U.S. Practice[2]

Commuter Rail / Transit Agency	Weekday Ridership (000s) (2)	Fare Media	Fare Options	Fare Structure	Peak Fares	Onboard with Surcharge	Onboard No Surcharge	Onboard Credit Card	No Onboard Sales	TVM	Ticket Window	Web/ Mail	3rd Party Vendor
No. New England Pass. Rail Auth.-Amtrak Downeaster - Portland, ME	1.5	P	S, M, T	D					x	x	x	x	
Music City Star - Nashville, TN	0.9	P	S, M, T	Z					x	x	x	x	x

Current State of the Practice in Commuter Rail

The table above shows the results of a scan of the practice for all commuter railroads listed in the APTA 2008 *Public Transportation Fact Book*, in addition to several other services that began operation after that report was compiled. It is important to note that the Alaska Railroad runs less frequently than the others and mainly serves seasonal tourists rather than commuters. Also, two of the services on the chart (PennDOT and NNPRA) are essentially state-funded supplemental Amtrak services, with a greater proportion of inter-regional travel than is common for most commuter railroads. Some of the commuter railroads, such as the Massachusetts Bay Transportation Authority and Metro-North, are part of larger multi-modal transit agencies, while others, like the Altamont Commuter Express and Virginia Railway Express, are managed as stand-alone rail services.

Paper-based tickets have long been used by commuter rails throughout the United States. Findings from this scan show that that they are still the standard fare media, with nearly every agency offering non-electronic single-ticket, multi-ride, and time-based passes (e.g., day passes, monthly passes). Even systems that have introduced more advanced fare-collection technologies have not completely transitioned over from paper-based media—monthly passes remain largely "flash" passes, which are visually inspected by conductors onboard, and handheld devices print paper ticket receipts.

All agencies other than the Alaska Railroad offer some form of time-based pass for unlimited travel. Seven of the agencies vary fares by peak- and off-peak service. All 24 agencies also offer discounted fares, ranging from free boarding for children under 11, to senior and student discounts. These discounts and distinctions are notable for the complexity that they introduce into fare-collection procedures and the design of any automated fare-collection system.

All 24 commuter railroads have distance-based or zone-based fare structures. Eleven systems use a distance-based structure, while the remaining 13 use zones. Flat fare structures, which are common in light rail systems, are not used by any U.S. commuter railroads.

There is a sharp divide in the ways in which commuter rails sell tickets and enforce payment. Thirteen of the railroads use a proof-of-payment system, which requires that tickets be purchased prior to boarding; there are no onboard sales, and riders face a fine or penalty if they board without a ticket. By contrast, the other 11 railroads allow onboard payment, though eight impose a surcharge if there was a TVM or ticket window available to the passenger before boarding. (SEPTA adds a surcharge even for passengers boarding at stations with no ticket purchase options.) Ticket sale outlets include ticket vending machines (17 agencies), ticket windows (16), advance online/mail order options (20), and third party vendors (6). Currently, only Central Puget Sound Regional Transit Authority and North County Transit District have plans to implement a smart card that will include their commuter rail transit option. Both of those systems plan to roll out their regionally integrated smart cards in spring 2009. Two agencies, New Mexico Rail Runner and Metro-North, are using handheld devices to facilitate onboard ticket sales. Only Rail Runner accepts credit cards onboard. While many railroads now offer online sales, Rail Runner is the only agency that allows customers to print out their own barcoded tickets in advance.

Spotlight on Innovation

As part of the Regional Rail On-board Electronic Payment Project, FTA, the Southeastern Pennsylvania Transportation Authority (SEPTA), New Jersey Transit, the Port Authority Transportation Company, and senior design project students from Temple University collaborated to develop a proof of concept for an electronic fare payment system for commuter rail services. The purpose of this project is to integrate several non-proprietary plug-and-play, off-the-shelf products and record onboard cash fare transaction and process electronic payments in the barrier-free environment of a commuter rail. These products include personal digital assistants (PDA) and wireless 802.11 access points, which can ensure easy, accurate, secure and efficient data and fare collections.

      Phase 1 of the project included a wireless-enabled handheld PDA to facilitate cash transactions and provide a connection to a central database. Phase 2 integrated enhanced PDA capabilities, including operability with magnetic media and contact smart cards, with plans to expand capabilities to read contactless smart cards in real time. Phase 2 also included a more sophisticated back-end database system. The system developed remains a proof of concept, due to changes in SEPTA's plans to upgrade fare collection technology from a closed system to an open-loop payment system; however, the graphical user interface on the PDAs may still be used in SEPTA's final smart card implementation. On June 12, 2008, the project team conducted a full system demonstration for SEPTA (including both the software and hardware), demonstrating its interoperability with SEPTA's back-end process.

3. Case Studies

3.1 Case Study Summaries

3.1.1 New Mexico Rail Runner Express

The New Mexico Rail Runner Express serves the metropolitan areas of Santa Fe and Albuquerque. Service is provided on a single rail line of approximately 95 miles, spanning six fare zones. Service to downtown Albuquerque started in July 2006, and an extension to Santa Fe opened in December 2008. Average weekday ridership is around 2,700, and about 50 percent of riders hold monthly passes.

Because the Rail Runner is a new system, the agency had the opportunity to leapfrog legacy systems to select a fare-collection system based on more advanced technologies, though they faced similar challenges to implementing AFC in a system without fare gates. Smart cards were not selected because Rail Runner, with its very small office staff, preferred to avoid the administrative complexity involved with becoming a card issuer and payment system manager. Instead, an innovative fare collection approach was developed using onboard and online sales and handheld devices, with no ticket vending machines (see sidebar). Abanco was selected as the vendor for fare collection and Herzog Transit Services provides operations services. Rail Runner staff expected that this approach would ultimately be more cost-effective than purchasing, maintaining, and servicing a full complement of TVMs, which were thought of as an older technology.

Once selected, the fare-collection system took about a month to implement and deploy, as the vendor had designed similar systems for airlines and other customers. Rail Runner did not encounter any major barriers to implementation, and the only unanticipated technical issue was that the handheld units require frequent battery recharging.

Passengers benefit from the choice between prepaid or onboard sales and the flexibility of being able to use debit/credit cards, even onboard. While initially light (15 to 20 percent), Internet sales now account for over 35 percent of total sales. Rail Runner prefers web sales, as they

Riding Rail Runner Express

Unlike most other transit systems, there are no ticket vending machines or ticket windows. Instead, Rail Runner riders can purchase their tickets in one of two ways: in advance from the agency's website or directly from the train's onboard staff using cash or credit/debit cards. Single tickets as well as daily, monthly, and annual passes are available through both sales channels, and all sales feed into the same financial database. Each ticket or pass product has a barcode that is validated by onboard customer service agents using a handheld scanner.

reduce the amount of cash involved and can be verified in real time. The agency's goal is to have internet sales reach over 70 percent, and as such, they have recently begun offering an online discount to encourage internet sales. For those customers who do buy onboard, the use of onboard ticket agents has the benefit of human interaction; the agents assist riders and provide information, which is particularly useful for a relatively new service which is in the process of building a ridership base.

Rail Runner's approach also reduces the complexity of its financial system. Handheld devices are brought into the office nightly and cleared into a central financial database, allowing both online and onboard sales to feed into the same financial system. Back-end costs are minimized since there is no need to service TVMs, and costs such as accounting, cash processing, and audits are minimized. Only one part-time accountant is required. Another benefit of the technology is that the source code is the same for the handheld device and online sales, making system updates easier.

One potential drawback of the system is that all barcodes must be scanned, including those on monthly passes, which increases the time needed for agents to inspect passes in comparison to a system with visual checks or a proof-of-payment approach. So far, this has not proven problematic. Also, onboard credit/debit transactions cannot be authorized in real time until wireless internet access becomes available in late 2009. To reduce the number of transactions that conductors make, Rail Runner explored the possibility of installing onboard TVMs which would accept credit/debit cards only. Though this has not been implemented, it is an option for the future, and sufficient electrical power is available in the train cars.

Regional fare integration has largely been pursued through low-tech means. Rail tickets and passes can be used as flashpasses for payment on multiple local bus operators as a result of bilateral agreements, with Rail Runner compensating the bus agencies for foregone revenue. Rail Runner staff report little pressing need for further integration or combined passes, though with the extension to Santa Fe they have been working on plans to offer a pass which combines the rail with connecting express bus service.

Some banks in the region have expressed interest in installing ATMs on the station platforms, and Rail Runner is exploring the options for using these ATMs as an additional means for advance ticket sales. Longer-term plans being explored include the potential use of cell phones to purchase and verify tickets, and the installation of online sales kiosks in conjunction with new real estate developments around stations.

3.1.2 Virginia Railway Express

Virginia Railway Express (VRE) has served the metropolitan Washington, D.C. area since 1992. VRE's system consists of two lines divided into nine fare zones. The system is heavily used and most trains are at capacity.

Many VRE commuters connect to the Washington Metropolitan Area Transit Authority (WMATA, or Metro) system and other regional services. For these riders, VRE offers a Transit Link Card which combines a VRE flash pass and a Metrorail magnetic stripe card. VRE also has an agreement with MARC, whereby both railroads honor each other's tickets for reverse commute through-trips via Union Station. By agreement with Amtrak, VRE riders holding multi-ride tickets or passes may also purchase "step up" tickets which are accepted on certain Amtrak services in the same corridor. These approaches to regional integration have generally worked well, but neither VRE nor MARC have yet found a viable way to join with WMATA's SmarTrip (contactless smart card) system, which would offer true integration on a single card. SmarTrip is not compatible with VRE's barrier-free environment. Moreover, due to reliance of a proof-of-payment system, using handheld devices to read SmarTrip onboard is not viewed as desirable, both because the technology investment would strain VRE's finances and (more importantly) because the additional transaction times involved with electronic readers would seriously overstretch their train crews during peak periods.

VRE is unique in that Federal employees comprise over half of its ridership, so coordination with the Washington area's Federal transit benefits program is an important consideration. Employee transit benefits typically take the form of transit vouchers (paper or electronic) or a pre-loaded SmarTrip card. There are a number of barriers to using these benefits for VRE tickets and passes: as noted above, VRE is not equipped with smart card readers, and even more basic transfer of revenue data between WMATA and VRE is hampered by use of different contractors and technologies for fare vending machines. The cost of

---

Riding Virginia Railway Express

Virginia state law requires VRE passengers to have a valid ticket before boarding. Most fares are purchased through credit/debit transactions and/or transit benefits, and most riders purchase ten-trip or monthly passes. Five-day passes and single tickets are also available.

Riders may purchase tickets online, from TVMs at stations, or through VRE's small vendor network at offsite locations and the Commuter Stores. Before boarding, riders must validate tickets on the platform. Most, but not all, passengers have their ticket or pass visually inspected. The policy is zero tolerance and the penalty is a minimum fine of $150 and a court summons for fare evasion.

most VRE monthly passes also exceeds the maximum transit benefit available,[3] meaning that some other form of payment is also needed to complete the transaction. Some workarounds have been developed via Commuter Direct, a local partnership that allows commuters to use their employee transit benefits to order passes for VRE (and other local systems) online and by mail.

In recent customer satisfaction surveys, VRE riders gave the agency relatively high grades (averaging around a "B") on the reliability of the ticket vending machines, ability to redeem transit subsidies, and overall ease of buying a ticket. However, VRE staff are aware of some customers' frustrations with the current approach and the potential benefits of becoming a full member of the SmarTrip regional smart card system. Due to the technical and institutional barriers noted above, VRE is taking a wait-and-see approach with no current plans to alter the fare collection system, which has been used since VRE's inception. Fare collection procedures and associated staffing levels are also written into VRE's contracts with their operating crews, so any changes would not be taken lightly. As new technologies emerge, some of VRE's key considerations would be the ability of conductors to check tickets in a timely fashion, even on crowded rush-hour trains, and the compatibility of the technology with other regional systems.

---

[3] Recent Federal legislation raised the tax-free limit on employer-provided transit benefits from $120 to $230 per month. As of this writing, the Federal government and most private employers have not yet adjusted their transit benefits. The costs of VRE monthly passes for the outermost fare zones (7-9) exceed even this new threshold.

3.1.3 Sounder Commuter Rail

Sound Transit is a regional transit agency for the Puget Sound area that operates a network of express buses, a light rail line, and a commuter rail line called the Sounder. The Sounder's North Corridor runs from Everett to Seattle and the South Corridor runs from Seattle to Tacoma. Starting in 2012, an extension will add service between Tacoma and Lakewood. The North Corridor operates four round-trip trains every weekday while the South Corridor operates eight round-trips every weekday. The system carried about 2 million riders in 2007.

The Central Puget Sound Regional Fare Coordination Project, an initiative to implement a coordinated regional fare structure and a single regional fare card, began in April 2003. In addition to Sound Transit, project partners included King County Metro, Pierce Transit, Community Transit, Everett Transit, Kitsap Transit, and Washington State Ferries. The goal of the project was to allow seamless travel and fare payment across the Puget Sound region regardless of which services and transit providers a particular trip might involve.

ERG Transit Systems was awarded the contract as the system vendor. After several years of planning, the agencies conducted a five-month revenue service "beta test" phase from 2006-2007 with 10,000 volunteers. The survey results showed mainly positive feedback and the agencies decided to move ahead with a full implementation. The contactless smart card was given the name ORCA, an acronym for "one regional card for all" and a reference to the region's marine life. The rollout was initially planned for the fall of 2008, but was postponed until late spring or early summer 2009 to allow more time for equipment installation and operator training. For the first six months, the ORCA card will be free with a minimum of a $5 purchase; afterwards it will cost an additional $5. Paper tickets will only be available for day passes and single rides.

The ORCA card can hold time-delimited passes and/or an "e-purse" (stored value debit account) on an embedded microchip, and is compatible across all seven participating transit agencies. With the exception of the Washington State Ferries, users will have full transferability of fares – that is, customers transferring between

---

Riding the Sounder

Riders must have proof of payment before boarding Sounder trains and can use cash, debit, or credit at TVMs on the platform to purchase tickets and bus transfers. The PugetPass, a regional discounted multimodal pass, is accepted on Sounder trains, but an upgrade ticket must be purchased at a TVM if the rail fare exceeds the pass value.

Starting in March 2009, a regional smart card, ORCA, will be fully rolled out to the public and will be sold from TVMs on the platform. Value can be added online, by mail, phone, or at participating retail outlets. Riders pay by tapping the ORCA card at a standalone fare transaction meter before boarding and tapping again at their arrival station. Failure to tap out results in deduction of the maximum fare. Riders transferring to another participating system from the Sounder using ORCA pay only the difference in fare for a two-hour period.

Customer service agents with handheld devices validate payment onboard through random inspection. Penalties for noncompliance start at $25.

services will receive credit for any fares already paid, and any transfer privileges will be worked out automatically. This is a major benefit to riders in this region, which has had a complex set of reciprocal transfer arrangements between agencies. The ORCA card also supports discounted fare distinctions such as senior or youth designations, but for security and privacy reasons will not carry any personal information on the card.

Implementing the smart card approach in Sounder's ungated commuter rail environment required significant technological investment, notably card readers along the platforms at each station (pictured at left) and handheld devices for conductors. Live network connectivity was added to the platform card readers to enable immediate feedback when a card is read for validity. The handheld devices do not yet have real-time connectivity but are synchronized overnight.

Sound Transit staff cited customer convenience and the creation of a truly regional system as the agency's chief motivation for participating in the ORCA project. Other anticipated benefits are improved ridership data and agency cost savings through reduced cash handling and streamlined back-office operations. One of the recurring challenges has been the difficulty of accommodating the fare structures from each of the agencies involved and ensuring that revenues are allocated fairly. At present, revenues from transfers and multi-agency passes are apportioned using a formula tied to a ridership survey conducted in 1994. A similar survey will be distributed right before ORCA's launch to establish a new baseline for revenue allocation, with the distribution among the participating transit agencies based on ridership and agreed-upon business rules. Each time a rider charges an ORCA card, the revenue collected will be automatically distributed to the correct agency according to this formula. Ridership data received directly from ORCA users will help to eliminate the need for paper surveys in the future.

One of the Sounder's chief concerns with ORCA is that using handheld devices will require more time for verification of fares than the use of flash passes, leaving conductors with insufficient time to conduct inspections. Sound Transit will be conducting time and motion studies on this topic and revisiting staffing guidelines in light of any changes brought about by ORCA. Other concerns include the potential for long lines of customers waiting to "tap out" at fare meters when large numbers of passengers disembark, as well as privacy issues surrounding the information stored on the smart cards. Sound Transit is researching data-protection measures in response to privacy concerns.

In the future, this system could be made interoperable with other transit agencies beyond the current seven partners. The system is also flexible enough to incorporate other applications such as parking payments, highway tolls, or even retail payments. At present, however, the focus is on transit payments only, as the agencies want to ensure that ORCA works well for its core mission before branching out into other areas. Employer transit subsidy programs would introduce some complexities in using the cards for anything other than transportation.

3.1.4 San Diego Coaster

North County Transit District (NCTD) and Metropolitan Transit System (MTS) operate the Coaster commuter rail in Southern California. In addition to the Coaster, other transit services in the region include bus, trolley, and the Sprinter light rail. Coaster's single line runs along the coast from Oceanside to San Diego, serving two main employment districts, Serrano Valley and downtown San Diego. The line has eight stations separated into four fare zones.

A contactless smart card called the Compass Card is expected be made available to the general public in the spring of 2009. The vendor for the smart card is Cubic Transportation Systems, Inc. The card will offer one method of payment and a coordinated tariff structure for all of the region's bus, trolley, light rail, and commuter rail services. Each card will be registered to a specific rider, enabling automatic reloads via an online account and discounted fares for seniors and the disabled. Current transit passes will also be replaced by the Compass Card, as the smart card is capable of holding passes and/or an e-purse (stored value account). Paper tickets will be phased out toward the end of 2009.

Two pilot tests were conducted to test the functionality of the Compass Card. The first pilot test involved about 1000 transit employees, and was used to test the basic readiness of the system and to help identify software glitches. In the second pilot test, NCTD and MTS opened up the test pool to 2500 members of the general public. In follow-up focus groups, one of the key findings was that riders disliked having to "tap out" at arrival. As a result, Coaster staff made a policy decision to make tapping out optional for monthly pass holders, since there is no need to debit their account for a specific fare.

The public pilot test also sparked curiosity from other riders who were unaware of the new Compass Card, leading NCTD to field questions from customers. One lesson learned is that the agency needs to be more

---

Riding the Coaster

Coaster requires proof of payment before boarding as there are no onboard sales. Compass Card, the region's new contactless smartcard, is replacing existing tickets and passes and will be sold online, by phone, at TVMs in certain transit stations, and at the transit store.

Fare validation devices have been installed on each station platform. Riders using the Compass Card "tap in" at the validation box upon boarding and "tap out" upon alighting, and the system debits the appropriate fare. (The maximum fare is charged to those who do not tap out.) Monthly pass holders do not need to tap out, though they are encouraged to do so to provide the agency with better ridership data.

Onboard, conductors use handheld devices to inspect the validity of Compass Cards and ensure that payment has been made.

proactive about informing the public; future public pilots of new technologies should be accompanied by marketing materials that will help explain the system, ideally with a handout that interested riders can take home with them.

Switching to the Compass Card and its tap-in/tap-out approach required some small adjustments to the fare structure. There was also some initial concern that the new fare collection system might present difficulties due to the extra time needed for onboard inspection, compared to quick visual checks. However, since Coaster is a proof-of-payment system, not all riders are inspected. Moreover, the length of the route and the spacing of the stations generally afford enough time for conductors to make their rounds. From the riders' perspective, they will only need to be concerned with remembering to tap in and possibly tap out to correctly charge their card under this new system.

The region's transit agencies have discussed incorporating paratransit into the Compass Card within the next few years. There is also a possibility of adding onboard sales to the rail lines.

3.1.5 Metropolitan Transit Authority New York - Metro-North

The Metro-North provides service between New York City and its northern suburbs in New York and Connecticut. The operating region east of the Hudson River comprises 120 stations spanning three lines, and serves, on average, over 270,000 weekday riders across nine counties. (Service west of the Hudson is operated by New Jersey Transit and is not included in the discussion that follows.)

In the late 1990s, Metro-North shifted from a manually operated system to a more automated one, and added TVMs as a part of that effort. More recently, in order to improve customer service while also increasing efficiency and reducing accounting processes, Metro-North introduced handheld devices for conductors, replacing the traditional "duplex" punch ticket that has been used for onboard sales.

The handheld devices, which use software written in-house (Windows Mobile®, Blu-ray®), are small enough to clip onto a conductor's belt. Ridership and fare information are downloaded into a central accounting database when the conductor docks the device. Among the benefits of the system are improved operating efficiency and more transparent accounting. Riders are pleased with the printed receipts which are easier to understand.

MNRR successfully tested the devices in the spring of 2008, and rolled out the system to all conductors in the following months. The start-up cost for the system, including the devices, software, new receipt stock, training, and a wireless contract, was $3.6 million. The handheld devices were purchased from Intermec, and the printers were purchased from Zebra. While the system has been successful, some limitations and challenges do exist. Initially there was some resistance among the conductors who did not want to change to a new practice. According to the *New York Times*, replacing the large conductor handbook with an electronic version on the PDA facilitated conductor acceptance. Metro-North is also working to extend the battery life of the devices.

Currently, the devices can only accept cash payment for a single-fare ride, and can only calculate a fare for a single origin-destination pair. Future management decisions could include upgrades to allow for credit and debit card acceptance.

> **Riding Metro-North:**
>
> Metro-North operates a conventional ticketing system, with a variety of tickets and passes available at TVMs, ticket windows, and online. Conductors also sell tickets onboard as they move through the cars checking passes and tickets. Over time, the railroad has moved toward a three-tiered tariff system. Riders pay the base fare when purchasing tickets at TVMs on the platform or at station ticket windows. A small discount is offered for internet sales, and a surcharge is imposed for tickets purchased onboard when a TVM or ticket window was available. Onboard sales are cash-only, while debit and credit cards are accepted for TVM and ticket window purchases.
>
> In 2008, Metro-North introduced electronic handheld devices to facilitate onboard purchases.

Regional integration efforts include a joint pass which combines a monthly commuter rail pass and a Metro Card, as well as a UniTicket which adds connecting service to a weekly or monthly rail ticket. MTA is currently in the early stages of conducting a study examining regional fare integration, which would include all agencies under the MTA as well as other local transport agencies.

3.1.6 Shore Line East

Shore Line East is a fully owned subsidiary of the Connecticut Department of Transportation. SLE provides daily service along a single line in southern Connecticut between New Haven and Old Saybrook. (In addition, one daily round-trip extends to New London, and some peak-hour weekday trains continue through New Haven to Bridgeport and Stamford.)

With a relatively small system (2300 daily riders) and eight- to ten-minute running times between stations, conductors generally have no trouble making it through the entire train to inspect passes and sell tickets. Overall, this conventional fare-collection system is viewed as working well, and SLE does not see any pressing need to shift to advanced or electronic fare-collection systems at this time. One acknowledged drawback of the current approach is the large amount of currency that conductors need to carry in order to make change for passengers buying tickets on the train. SLE staff are working to address this issue and are considering the possibility of imposing a surcharge on onboard purchases during times when a ticket window is available. Another area of discussion is payment enforcement. At present, SLE handles this with a light touch, particularly for regular customers who have simply misplaced their passes, but a more formal approach may be needed as ridership grows.

Shore Line East riders often connect with other transit services, making regional integration an important issue. This is particularly true with respect to the Metro-North Railroad, since SLE estimates that 40-50 percent of its ridership continues westward beyond New Haven. At present, this integration is handled using UniRail tickets, which are Metro-North monthly passes with a special SLE endorsement. These tickets are produced by Metro-North by special agreement. For SLE customers connecting in New Haven, Metro-North also waives its normal surcharge for onboard ticket purchases. SLE offers a monthly pass (printed on Amtrak cardstock) which combines SLE rail with connecting shuttle buses in downtown New Haven. There is also an arrangement in place with Amtrak for acceptance of SLE ten-trips and monthly passes on certain Amtrak regional trains operating along the same line. These relatively low-tech solutions have proven adequate for current needs. However, SLE staff indicated that they may re-evaluate their fare-collection approach if commuter rail service is added in the New Haven-Hartford corridor.

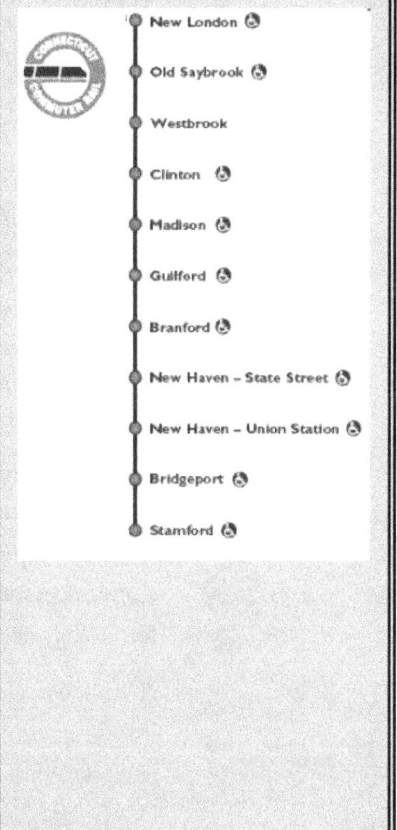

Riding Shore Line East:

Riders may purchase tickets, 10-trips, and monthly passes at ticket windows at major stations (New Haven, New London, and Old Saybrook) or by mail or phone. One-way and round-trip tickets may also be purchased onboard with cash. There is no surcharge for onboard purchases. Amtrak crews operate and staff the service.

Improved ability to count and identify passengers in case of emergency was cited as a potential benefit of some forms of electronic fare collection.

3.2 Lessons Learned

The six case studies above provide some insight into commuter railroads' experiences in adoption of AFC. The lessons learned summarized below are intended to be of use to agencies considering a move to AFC. For proposed commuter rail systems that are still in the planning stages and can thus implement a new fare-collection system from scratch, this information can also help to identify the pros and cons of different approaches.

- Unsurprisingly, the barrier-free environment of commuter railroads and their complex fare structures require AFC solutions which are different from those of gated urban transit systems. Using a proof-of-payment approach (versus conventional fare collection with some onboard sales) also has implications for the design of an AFC system.

- Specifically, agencies must consider the impact of fare collection technologies on conductors' workloads (all), fare policy (Coaster), transit benefit programs (VRE), and possibly even train car and station design (Rail Runner—electricity & Wi-Fi on trains; Coaster and Sounder—fare validation boxes on platforms).

- Agencies that have adopted AFC generally anticipate benefits in the area of customer satisfaction and convenience, better regional integration of multiple modes and services, and reduced accounting and back-office costs.

- To facilitate adoption, pilot testing and user feedback are important and can help identify areas where technologies or policies need adjusting (Coaster, Metro-North).

- New railroads, as in the case of the Rail Runner, have the opportunity to approach fare collection in innovative new ways.

- Most agencies still favor a non-electronic multi-use ticket or pass which allows for travel on multiple modes of transport and can even function across different agency services. In particular, for smaller agencies with modest ridership and relatively simple operations, the benefits of AFC may not yet outweigh the costs of new technology and equipment (Shore Line East).

- Adopting a regional smart card requires financial backing and upgraded physical infrastructure. Having many different transit providers in a region can spur innovation and cooperation (e.g., Sounder and Coaster) *or* create barriers to further integration (VRE). Regions with many smaller agencies (e.g., Puget Sound) require significant coordination and cooperation to implement a regional smart card.

- The growing interest in and acceptance of financial industry cards, employer identification cards, mobile payment devices and other account-based payment methods may provide additional opportunities for commuter rail fare payment.

4. Domestic and International Experience with Automated Fare Collection

Additional insight on electronic fare collection for transit and some indications of future directions can be found in the experiences of other agencies around the world, though most innovations have been in urban transit rather than in commuter rail. This section discusses two main approaches: "closed-loop" systems, which are proprietary payment networks that are typically built and administered by the transit agency; and "open-loop" systems, in which bank cards are used to pay fares directly and processing is handled through the international banking system rather than the transit agency's own network.

4.1 Closed-Loop (Proprietary) Systems

Transport for London has been promoting the Oyster card, with regional interoperability ever since 2003. Today, over ten million cards have been issued and the smart card can be used on the bus, tram, Tube, Docklands Light Railway (DLR), London Overground, and certain National Rail services. The electronic payment system utilizes Near Field Communication (NFC) technology and operates under the Mifare technical standard. NFC is a short-range wireless connectivity technology which enables communication between two NFC-compatible devices when they are brought within four centimeters of one another. Compatible devices range from mobile phones to digital cameras. After an NFC connection is established, other compatible communications technologies such as Bluetooth or Wi-Fi can be used. Oyster cards can hold up to £90 of stored credit on a pay-as-you-go plan and/or hold up to three season tickets in the form of a bus pass, tram pass, or Travelcard. The Oyster card guarantees daily price capping, such that riders never pay more than the maximum daily rate for their itinerary. Riders may purchase pay-as-you-go fares, but pay a refundable £3 deposit to obtain the Oyster card for the first time. To use the card, riders must tap in and out. Failure to do so when needed results in a charge of the maximum fare. This smart card payment system uses Mifare technology and is operated by TranSys.

Since Hong Kong launched the Octopus card in September 1997, over 17 million cards have been distributed with a 95 percent penetration rate in people of ages 16-65. Due to its popularity, this smart card comes in various physical forms as creative as children's wristwatches and ownership can be designated as child, student, adult, or elder, or even customized. Octopus card technology includes an embedded RFID chip. The cards can be used on Mass Transit Railway (MTR), Kowloon-Canton Railway (KCR), Star Ferry, light rail, trams, and CityBus Limited. The Octopus cards are also accepted at grocery stores, convenience stores, fast food restaurants, retail outlets, and parking meters, and are even used to record school attendance and for building access control. Card holders can earn reward dollars for using their Octopus card at retailers and obtain discounts on railway fares. This smart card payment system was designed by the Australian ERG Group.

Anyone may purchase or "lease" an Octopus card at transportation customer service centers. The on-loan cards are generally for residents and require a deposit and application to obtain. Non-refundable cards do not require a deposit .The cards can be linked to a credit card from any one of 22 participating banks for automatic refills or value can be added at vending machines, vendors, and ticket offices. The card's balance can actually be negative (up to 35HKD) before it will be rejected as invalid; this gives riders a chance to complete their journey and refill their card at a later time.

The Octopus card has expanded beyond Hong Kong into two cities in China, Shenzhen and Macau, as of 2006. So far, use in these cities is more limited—the card is only accepted at a few cafes and fast food restaurants.

Los Angeles County Metropolitan Transportation Authority's Transit Access Pass (TAP) combines a contactless Visa card (payWave) with a stored-value transit account. The cards are accepted on LACMTA services and at any location which accepts Visa, with transactions drawn separately out of the two different "purses." Funds can be transferred from one purse to the other. Riders may also purchase a basic card, which can store up to $500, at ReadySTATION kiosks at the LA Metro stations. The more personalized card, aimed at riders without bank accounts, holds up to $10,000, and allows for ATM withdrawals and direct deposits. These can be obtained by mail or phone.

Transport for London is seeking to reduce fare-collection costs by transitioning to open-loop bankcard payment, with a possible pilot test in 2010. Currently, the co-branded OnePulse card issued by Barclay carries the proprietary Oyster purse. Octopus Cards Ltd. is unlikely to allow competing open-loop payment, but allows Octopus to ride on Citibank co-branded cards.

4.2 Open-Loop Payment Systems

The New York City subway trial of MasterCard's PayPass began in 2006 and now has 30 participating stations. Although some lessons may be gleaned from this trial, it is not directly applicable to commuter rails, as the New York Metro is a closed, gated system. Any Metro passenger with either a Citibank credit or debit MasterCard with PayPass can use their card, tag, or mobile phone to purchase subway fares. The embedded computer chip enables a safe, contactless payment at the turnstile. As an incentive to use the PayPass system, a rider who prepays for 12 trips through an online account can receive them for the price of ten. Online account features also include an auto-reload function, rider history information, and a billing statement.

As of November 2008, Visa implemented pilot tests for contactless transit payments in Paris and Los Angeles. Rather than having the individual transit agencies handle fare collection, both of these systems use Visa's payWave technology to eliminate the need for cash or ticket purchasing. Other transit pilot tests with Visa have taken place in South Korea, Kuala Lumpur, Hong Kong, India, Turkey, and UK. These locations use a variety of prepaid Visa cards, Visa payWave cards, or dual-use cards with Visa payWave and transit properties, depending on the level of integration between Visa and the local transit agency.

Regie Autonome des Transports Parisiens (RATP) in Paris instituted a smart card, the Navigo pass, as of 2001. For the pilot test, a special RATP fare gate was set up to demonstrate how easily the Visa payWave cards could be integrated into the current system. RATP is now working with Visa Europe and MasterCard to bring open-loop payment to RATP gates and buses in order to more efficiently serve the large number of visitors using separate magnetic-stripe paper tickets.

Other cities are also looking to open-loop bankcard payment as a possible fare-collection scheme. The Utah Transit Authority began accepting credit and debit cards from the four major U.S. brands for payments on its bus, tram, and commuter rail network in Salt Lake City, and is beginning a pilot test to accept Department of Defense Common Access Card smart card IDs for fare payment. According to *Cards and Payments,* Chicago Transit Authority hopes to go entirely to open-loop bankcard payment in four to five years, but is counting on funding from the banking industry for infrastructure. Philadelphia released a request for proposals in November 2008, seeking partnership for open-loop payment.

4.3 Future Outlook

Other fare-payment media and technologies are starting to emerge. Commuter railroads are already exploring possible future changes to their fare-collection systems, such as wireless connections onboard, dispensing fare media through ATMs (Rail Runner), increased acceptance of credit/debit cards onboard, and scanning barcodes on mobile phones. As new technologies emerge, whether in commuter rail or other types of transportation, ongoing study will help provide a sense of their performance and suitability for the commuter rail environment.

Mobile Payment Technologies

The development of mobile payment for transit applications has been slowly gaining traction around the world. Benefits of using phones for payment include the elimination of cash and vending machines, customer convenience, and faster travel. The domestic and international examples listed below provide evidence that these technologies can be successful in the real world and instrumental as payment methods of the future.

In January 2008, Bay Area Rapid Transit (BART) and Jack in the Box, a fast food chain, launched a joint NFC field pilot for fare and fast food payment through a mobile phone. This pilot is the first to integrate transit payments and mobile phones in the United States. It operates through a prepaid electronic purse carried on a Sprint mobile phone, which can be topped up through an over-the-air feature. The trial included 230 participants who took 9000 BART trips over the four-month time period of the trial. Participants could also tap their phones on smart BART advertisements at transit stations. Public response favored this experimental system, as 80 percent of the participants claimed the system was easy to use and convenient.

After three years of field testing, China Unicom and Yucheng Transportation Card have commercially launched a NFC mobile payment method called "Cqpass" as of January 2009. This new mobile payment replaces the Yucheng Tong Card and can be used at restaurants, bus and cable car stations, hotels, and other participating retailers. Phase 2 of this roll-out will incorporate a bank account for automatic top-ups.

In October 2007, EZ Link and StarHub, a mobile service provider in Singapore, conducted a six-month trial to embed the EZ Link transit card into a mobile phone. This trial is one of the largest existing applications, as there are over 20,000 EZ Link acceptance terminals. Like the BART trial, mobile phones can also be tapped on smart posters to receive promotional offers or service updates. The post-survey results reveal strong interest in a permanent mobile installation from consumers, but it may take awhile for Singapore to develop the infrastructure necessary to support a full-scale implementation.

Transit riders in Helsinki, Finland can purchase a single ride ticket good for the metro, ferry, bus, tram, or certain commuter trains via their mobile phones. The ticket will arrive as a text message to the phone and include an identification number and time validation. These ticket purchases will be included as part of the phone bill statement. For non-single-ride tickets such as time-based or trip-based tickets, commuters will still have to use a Travelcard, which is the area's regionally integrated smart card.

As of March 2009, Czech operator Telefonica O2 and the transport municipality of Pilsen in the Czech Republic will launch a nine-month NFC mobile payment trial. This trial will utilize Nokia phones and Mifare technology. During the trial's first phase, until June 2009, participants can download the NFC transit application. By the second phase, they will be able to top up and buy bus tickets directly from their phone.

5. Conclusion

Urban transit agencies in the United States and abroad have moved toward various forms of EPS and AFC over the past two decades. Although adopters have been rewarded with a number of benefits, adoption on commuter rail systems has proven challenging and has lagged behind that of other transit modes.

As Phase I of an envisioned two-phase effort, this report provided a general overview of the relatively limited experience to date with EPS on U.S. commuter railroads, including key lessons learned and analysis of barriers to broader adoption.

Paper-based tickets have long been used by commuter railroads throughout the United States. Findings from this scan show that that they are still the standard fare media, with nearly every agency offering non-electronic single- and multi-ride tickets and passes. Even systems that have introduced more advanced fare-collection technologies have not completely transitioned over from paper-based media—monthly passes are still primarily "flash" passes which are visually inspected by conductors onboard, and handheld devices print paper ticket receipts.

Currently, only two systems, Central Puget Sound Regional Transit Authority and North County Transit District, have plans to implement a smart card that will include their commuter rail transit option. Both of those systems plan to roll out their regionally integrated smart cards in spring 2009. Two agencies, New Mexico Rail Runner and Metro-North, are using handheld devices to facilitate onboard ticket sales. Only Rail Runner accepts credit cards onboard. While many railroads now offer online sales, Rail Runner is the only agency that allows customers to print out their own barcoded tickets in advance.

The six case studies in this report reflect more in-depth findings and showcase a variety of technologies, stages of adoption, and levels of regional integration. Lessons learned from the case studies include:

- The barrier-free environment of commuter railroads and their often complex fare structures require AFC solutions which are different from those of gated urban transit systems. Using a proof-of-payment approach (versus conventional fare collection with some onboard sales) also has implications for the design of an AFC system.

- Specifically, agencies must consider the impact of fare collection technologies on conductors' workloads (all), fare policy (Coaster), transit benefit programs (VRE), and possibly even train car and station design (Rail Runner—electricity & Wi-Fi on trains; Coaster and Sounder—fare validation boxes on platforms).

- Agencies that have adopted AFC generally anticipate benefits in the area of customer satisfaction and convenience, better regional integration of multiple modes and services, and reduced accounting and back-office costs.

- To facilitate adoption, pilot testing and user feedback are important and can help identify areas where technologies or policies need adjusting (Coaster, Metro-North).

- New railroads, as in the case of the Rail Runner, have the opportunity to approach fare collection in innovative new ways.

- Most agencies still favor a non-electronic multi-use ticket or pass which allows for travel on multiple modes of transport and can even function across different agency services. In particular, for smaller agencies with modest ridership and relatively simple operations, the benefits of AFC may not yet outweigh the costs of new technology and equipment (Shore Line East).

- Adopting a regional smart card requires financial backing and upgraded physical infrastructure. Having many different transit providers in a region can spur innovation and cooperation (e.g., Sounder and Coaster) *or* create barriers to further integration (VRE). Regions with many smaller agencies (e.g., Puget Sound) require significant coordination and cooperation to implement a regional smart card.

- The growing interest in and acceptance of financial industry cards, employer identification cards, mobile payment devices and other account-based payment methods may provide additional opportunities for commuter rail fare payment.

Lessons on electronic fare collection for all types of transit can be gathered from experiences across types of transit and across national boundaries. London and Hong Kong both have years of experience with successful, proprietary closed-loop systems and co-branded cards. London is now looking to expand to an open-loop system. New York, Paris, Malaysia, India, and Turkey have also conducted pilot tests. Other cities, including Chicago and Philadelphia are also looking to open-loop bankcard payment as a possible fare-collection scheme.

As new technologies emerge, whether in commuter rail or other types of transportation, ongoing study will help provide a sense of their performance and suitability for the commuter rail environment.

Areas for Further Study

Phase II of this study, if pursued, will build on these results with more in-depth analysis of the financial, management, and technology issues associated with specific AFC deployments. Depending on the specific AFC deployments selected for study, areas for further research would include some or all of the following:

- "Time and motion"-type study or case studies, related to the time requirements for various forms of EPS versus manual collection, particularly for conductors' time during peak periods

- Analysis of changes to back-office and accounting processes and quantification of any savings from EPS

- Processes and formulas for revenue-sharing among agencies in a regional fare-card partnership

- Effects of AFC on customer satisfaction, intermodal connectivity, and ridership

- Examination of the benefits and costs of open-loop and closed-loop systems

Bibliography

Anders, Marjorie and Brucker, Dan. "MTA Metro-North Railroad introduces hand held ticket machines for use on board trains to improve customer service." Metro-North Railroad Press Release, Metropolitan Transportation Authority, State of New York, July 9, 2008.

2008 Public Transportation Fact Book, 59th edition. American Public Transportation Association. Washington, D.C., June 2008.

Balaban, Dan. "Open-Loop Transit Picks up Speed." *Cards & Payments*, Vol. 22 No. 1, January 2009.

Belson, Ken. "With new device, Metro-North moves toward cashless ticketing." *New York Times,* July 10, 2008.

Cowen, Mike. "Pay pass 'pay as you go' development over the last 12 months." Mastercard Worldwide. http://www.moving-on-conference.co.uk/pdf/presentations/Mon/MikeCowen.pdf, June 6, 2008.

EZ Link. http://www.ezlink.com.sg/. Accessed February 2009.

Fare Policies, Structures, and Technologies: Update. Transit Cooperative Research Program Report 94. Washington D.C.: Transportation Research Board of the National Academies, 2003.

"International Micropayment." GBDe 2007 Issue Group. http://www.gbd-e.org/ig/imp/IMP_2007_Tokyo.pdf, November 8, 2007.

"Mastercard and RATP to cooperate to study the feasibility of accessing public transport systems." Smart Card Trends. http://www.smartcardstrends.com/det_atc.php?idu=8186&PHPSESSID=e8bacea31d8e22f313da93a0b78c0645, November 4, 2008.

MasterCard PayPass. http://www.mastercard.com/us/personal/en/aboutourcards/paypass/. Accessed February 2009.

Octopus. http://www.octopuscards.com/consumer/en/index.jsp. Accessed February 2009.

"Public transport in the Helsinki region." Helsinki City Transport. http://www.connekt.nl/www/filelib/userfiles/file/Studiereizen/HTC_general_Vanhanen_080918.pdf, September 19, 2008.

"What is Oyster?" Transport for London. http://www.tfl.gov.uk/tickets/oysteronline/2732.aspx. Accessed February 2009.

"Smart Card – signing of the interlocal cooperation agreement and vendor contract." http://transit.metrokc.gov/prog/smartcard/sc_signing042903.html, April 23, 2003.

Sounder Commuter Rail. http://www.soundtransit.org/Documents/pdf/projects/sounder/FACT_Sounder.pdf. Accessed April 2008.

"Transit operators in Paris, Los Angeles looking at Visa payWave." ContactlessNews. http://www.contactlessnews.com/2008/11/05/transit-operators-in-paris-los-angeles-looking-at-visa-paywave, November 5, 2008.

"Visa to improve payment experience for commuters in Los Angeles and Paris; working with transit operators to enable Visa Payment at the fare gate." Visa Press Release. http://www.corporate.visa.com/md/nr/press871.jsp, November 4, 2008.

"Welcome to Compass Card." San Diego Association of Governments. http://compass.511sd.com/. Accessed February 2009.

Photo Credits

1. p. 10—Photographs Courtesy of New Mexico Rail Runner Express
2. p. 12—Photo Courtesy of VRE
3. p. 15—Photo Credit: Lydia Rainville
4. p. 16—Photographs Courtesy of Sound Transit
5. p. 17—Photo Credit: Dan Egger
6. p. 21—Image Courtesy of Shore Line East, ConnDOT

www.ingramcontent.com/pod-product-compliance
Lightning Source LLC
Chambersburg PA
CBHW081758170526
45167CB00008B/3233